BEAT HIGH GAS PRICES NOW!

Also by Diane MacEachern

Save Our Planet:
750 Everyday Ways You Can Help Clean Up the Earth

Enough Is Enough – The Hellraiser's Guide to
Community Activism:
How to Organize a Successful Campaign for Change

www.theworldwomenwant.com

BEAT HIGH GAS PRICES NOW!

*The Fastest, Easiest Ways to
Save $20-$50 Every Month on Gasoline*

By Diane MacEachern

**Andrews McMeel
Publishing**

Kansas City

Beat High Gas Prices Now! copyright © 2005 by Diane MacEachern. All rights reserved. Printed in the U.S.A. No part of this book may be used or reproduced in any manner whatsoever without written permission except in the case of reprints in the context of reviews. For information, write Andrews McMeel Publishing, an Andrews McMeel Universal company, 4520 Main Street, Kansas City, Missouri 64111.

05 06 07 08 09 10 BID 10 9 8 7 6 5 4 3 2 1

ISBN-13: 978-0-7407-6002-0
ISBN-10: 0-7407-6002-5

Library of Congress Control Number: 2005935606

www.andrewsmcmeel.com

ATTENTION: SCHOOLS AND BUSINESSES
Andrews McMeel books are available at quantity discounts with bulk purchase for educational, business, or sales promotional use. For information, please write to: Special Sales Department, Andrews McMeel Publishing, 4520 Main Street, Kansas City, Missouri 64111.

To Dick, Daniel, and Dana

BEAT HIGH GAS PRICES NOW!

What's Inside:

- How to use the Internet to save money at the pump
- Why the time you shop could affect how much gasoline you buy
- Why where you shop could determine how much money you spend on gas
- How to maintain a vehicle's proper tire pressure to spend less on gas
- Easy ways to save gas by avoiding certain types of roads
- When it makes sense to use expensive "high-octane" gas – and when it doesn't
- How to use credit cards to get money back when you buy gas
- Why driving 10 mph slower on the highway can make every dollar spent on fuel go farther
- How to get a tax deduction for driving a more fuel-efficient vehicle
- Why tailgating wastes gas
- How to solve the gas-guzzling dilemma of the two-car family
- What you need to know when buying a new or used car
- What tricks make gas go farther if you're taking a driving vacation

and much, much more.

Contents

Introduction x

How Did Gas Get So Expensive Anyway? 1

Saving Gas Will Protect More Than Your Pocketbook
– It's Good for the Environment, Too 3

Why Drilling for Oil in the Arctic Is Not the
Answer 5

The Six D's 8

    DRIVE SMART 10

    DRIVE CHEAP 15

    DRIVE LESS 19

    DRIVE IN TUNE 25

    DRIVE A GAS STRETCHER,
    NOT A GAS GUZZLER 30

    DON'T DRIVE ALONE 33

Non-Gasoline Fuel Options 35

Saving Gas on the Job 38

Vacation and Traveling Tips 41

What to Consider If You're Buying
a New Vehicle 44

Try a Hybrid 48

More Car-Buying Tips 51

For More Information 53

The Top Ten Ways to Beat the
High Price of Gas... 56

About the Author.. 59

INTRODUCTION

If the price of gas is sending you scurrying for ways to "beat the pump," you're not alone!

From Seattle, Washington, to Silver Spring, Maryland, drivers used to spending $30 a week to fill up a 15-gallon tank are now spending upward of $45, $50, or more – and cutting into their already stretched home budgets to do so.

Things are just as bad in Canada, where pump prices have reached as high as $1.24 per litre (the equivalent of almost $5 a gallon).

Hurricane Katrina showed how directly America's dependence on oil can impact our pocketbooks. In that storm's aftermath, which damaged 10 percent of the nation's oil refining capacity, prices at the pump surged 50 to 75 cents a gallon in just two days in some parts of the country.

Growing worldwide oil demand, increasing terrorism, and the ongoing conflict in Iraq throw the future price of oil and gasoline under an even greater cloud.

Given such uncertainty, it makes great sense to prepare for gas prices to continue to rise. What can you do as a consumer to spend less money on gas and still get to where you want to go?

Lots – if you read *Beat High Gas Prices Now!*, a handy guidebook that offers dozens of tips to help you reduce the amount of money you spend on gas every week.

Here's what you'll find inside:

- Ways to reduce the amount of fuel you use when driving to work or just doing errands around town

- Information to help you choose a new car if you want to trade in your current gas guzzler for a gas stretcher

- Profiles of the new hybrids that get over 50 miles for every gallon of gas you put in their tank

- Suggestions for minimizing your fuel intake while on vacation

- Internet resources that will help you find the cheapest gas in your neighborhood, alternative fuels, carpools, and more.

I explain why drilling for oil in the Arctic National Wildlife Refuge – an extremely bad idea that keeps cropping up on Capitol Hill – is not the solution either to our energy crisis or to the high prices for gasoline we're currently paying at the pump.

As an added bonus, at the very end of the book I include a summary of the top ten ways you can save gas – and money – right now.

According to a recent survey by the nationally recognized Gallup polling organization, people expect current gas price increases to be permanent. If that's the case, we may as well be prepared. This isn't a crisis. It's an opportunity – to learn how to use gas more wisely and save more money doing so.

Let's get started. Beat high gas prices now.

How Did Gas Get So Expensive Anyway?

(And How Much More Are Gas Prices Going to Rise?)

One of the reasons why gas costs so much is because we all use so much of it. Even though the U.S. only has about 3 percent of the world's oil reserves, we consume 25 percent of the amount of petroleum that's produced worldwide. When it comes to supply and demand, our demand way outstrips our supply. Put another way, the more we use, the more it costs us.

Gasoline consumption has especially increased since Americans started driving sport utility vehicles (SUVs). In fact, according to the *Washington Post,* U.S. citizens use 24 percent more gas today than we did in 1990, thanks to the 84 million SUVs Americans are driving these days.

Natural disasters like Hurricane Katrina reflect how vulnerable we are to our dependence on oil. In what seems like the blink of an eye, Hurricane Katrina knocked out nine major oil refineries in the Gulf of Mexico and damaged four others. Fully 10 percent of the nation's ability to produce gasoline was affected.

The fact that the U.S. buys more than 60 percent of its crude oil from other countries, especially the Middle East, doesn't help either. When those countries limit how much oil they produce, the price we pay for gasoline usually goes up. If Middle East leaders decide to produce less oil, gas prices will rise even more.

How much more expensive is gas today than it was a year ago? Gas is made from crude oil. In 2004 crude oil only cost around $35 a barrel. One year later, crude oil prices have doubled, to over $70 a barrel, forcing gasoline prices to jump as much as a dollar a gallon over last year.

No one knows how high gas prices are going to soar. In Canada and Europe, consumers are paying close to the equivalent of $5 and $6 a gallon. Will prices in the U.S. reach such a level?

It's hard to say. Meanwhile, it makes sense to use fuel efficiently – and not only because it will save money at the pump.

Saving Gas Will Protect More Than Your Pocketbook – It's Good for the Environment, Too

Burning gasoline creates a whole host of environmental problems – from air pollution and smog to global warming and climate change.

Drilling and transporting oil destroys wilderness and wildlife areas. Refining petroleum generates a wide variety of toxic substances and can create hazardous waste sites. Using fuel efficiently helps protect air, water, and wildlife and improve the quality of our lives overall.

Here's just one specific example of how using less oil and gas protects the environment.

Vehicles that get fewer miles per gallon create more carbon dioxide, which is a primary cause of global warming. In fact, every gallon of gasoline your vehicle burns puts 20 pounds of carbon dioxide into the atmosphere. Fortunately, when you drive a fuel-efficient car, you reduce your impact on global warming. If you drive a vehicle that achieves 25 mpg rather than 20 mpg, you actually prevent the release of about 15 tons of greenhouse gas pollution over the lifetime of your vehicle. The potential carbon dioxide reduction for a car that gets 32 miles per gallon is 5,600 pounds per year.

So, using less gas is not just better for your pocketbook. It's better for our planet, too.

WHY DRILLING FOR OIL IN THE ARCTIC IS NOT THE ANSWER

Whenever any kind of gas crisis or oil shortage occurs in the United States, many politicians immediately turn their gaze toward the coastal plain of the Arctic National Wildlife Refuge in Alaska.

This pristine wilderness is home to 129,000 caribou, polar bears, musk oxen, wolves, foxes, and countless species of birds. The native Gwich'n people have lived here for generations as well, maintaining a unique lifestyle and culture that have become renowned the world over.

The coastal plain comprises the last 5 percent of Alaska's North Slope that is not already open to oil exploration and drilling. The United States Geological Survey is not sure there is even any oil under this wilderness – but if there is, USGS scientists estimate that it is very likely only enough to meet America's petroleum needs for six months. Oil executives themselves admit that it would take at least ten years to get the oil out of the ground, refined, and into a gas pump.

In other words, any oil that might come out of the Arctic Refuge would do little to lower the price of gasoline for consumers or reduce U.S. dependence on imported oil. What it would do is destroy a unique wilderness and wildlife habitat that most certainly would never recover.

New gas-saving technologies like the ones you'll read about in this book can save far more oil than would ever be produced in the Arctic National Wildlife Refuge, and at far less cost. For example, just improving the fuel efficiency of cars and light trucks by 0.4 mpg would save substantially more than the 156,000 barrels of gasoline that might be generated by Arctic Refuge oilfields. Simply replacing worn-out tires with tires that achieve the same efficiency as the ones the vehicle originally came with would save several Arctic oil fields' worth of oil.

When you hear about proposals to drill for oil in the Arctic National Wildlife Refuge, think about how much more practical, cost-effective, and environmentally smart it is to save energy at the pump.

You can get more information about the Arctic National Wildlife Refuge from the Alaska Wilderness League, www.alaskawild.org.

THE SIX D'S

No matter how much gas costs, it will always make sense to use as little of it as possible to travel the distance you need to go. After all, **the point is to reach your destination, not to spend your hard-earned money on gasoline, right?**

Using the tips suggested in this book you can **save $20-$50 every month** on fuel by driving smarter, driving more cheaply, driving less, and driving a vehicle that's well tuned up. If you drive a car that stretches the most miles out of every gallon of gas, your savings can increase even more. If you share a ride with others, just watch your gasoline savings mount.

Beat High Gas Prices Now! offers easy-to-remember tips that will help you economize at the pump. Once again, here are the six gas-saving strategies they fall into:

1. **DRIVE SMART**

2. **DRIVE CHEAP**

3. **DRIVE LESS**

4. **DRIVE IN TUNE**

5. **DRIVE A GAS STRETCHER, NOT A GAS GUZZLER**

6. **DON'T DRIVE ALONE**

⨯1⨯
DRIVE SMART

World renowned race car driver Jackie Stewart once said the smartest way to drive was as if there were an egg under the gas pedal. In other words, drivers should accelerate gently and evenly, stay with traffic instead of hopping like a jack rabbit in and out of lanes, and drive at a reasonable speed. What else can you do to save gas and money at the pump?

- **Drive sensibly** – Speeding, rapid acceleration, and frequent braking waste gas. Such an aggressive driving style can lower your gas mileage by 33 percent on the highway and by 5 percent around town, let alone pose a hazard to other drivers. So take it easy – save gas, save money, and be safe, too.

- **Drive the speed limit** – According to the U.S. Department of Energy, gas mileage decreases rapidly at speeds above 60 mph. In fact, every 5 mph you drive above 60 mph is like paying an additional $0.10 per gallon for gas. So stick to 60 or less and put the savings in the bank.

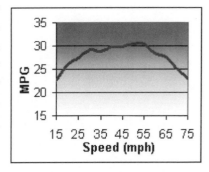

- **Use overdrive gears** – Overdrive reduces the car's engine speed, decreasing fuel consumption and limiting engine wear.

- **Cruise, and you won't lose** – Using cruise control on the highway helps maintain a constant speed that usually will help you improve your fuel economy by 4-14 percent.

- **Sometimes, just turn off the car** – Idling gets 0 miles per gallon. If that's not a waste of gas and money, I don't know what is! Cars with larger engines usually waste more gas idling than do cars with smaller engines; drivers of idling SUVs may as well just throw money out the window. If you find yourself stopped in a long line of traffic, at a railroad crossing, or waiting for kids coming out of school or an after-school activity, turn off the car until you can move along.

- **Avoid drive-through destinations** – Skip drive-through restaurants, banks, pharmacies, and other places of business that might require you to wait in long lines with your engine idling away the money you might otherwise spend on videos, clothes, or a computer upgrade.

- **Shift into neutral** – Even for very short waits, shift your vehicle into neutral to avoid burning gas unnecessarily.

- **Don't get lost** – Make sure you know where you are going before you begin your trip to avoid emptying the gas tank in search of a destination you can't find. Call ahead for directions, plot your course on a map, or use www.mapquest.com to get a set of directions you can print from the Internet.

- **Call ahead** – If you're going to a take-out restaurant or service, try calling ahead so your order is ready when you arrive to pick it up. Then turn off your engine, park your vehicle, and run in for your goods.

- **And about that tailgating . . .** – Don't hug the bumper of the driver in front of you. Looking 10 to 12 seconds down the road gives you time to anticipate circumstances and drive as efficiently as possible given current road conditions. Tailgating may force you to stop and start frequently, costing you 1-2 percent in fuel efficiency.

- **Know when to close the windows** – At highway speeds, it's more fuel efficient to use internal venting or air conditioning than to drive with the windows down, which can actually drag down your vehicle. At slower speeds, such as in the city, keep the windows down and avoid air conditioning, which uses more gas.

- **Turn the radio off first** – When your drive is at an end, turn off the radio, CD player, air conditioner and other power-consuming accessories before turning off the ignition. You'll minimize engine load the next time you start up.

- **Roll to a stop** – Slowly decelerating rather than abruptly braking is one of the best fuel-saving tricks you have up your driver's sleeve. When you see a stop sign up ahead or notice the car in front of you hitting its brake lights, take your foot off the gas and start coasting to a stop.

- **Keep calm** – Revving the engine just before turning off the ignition requires extra gas and may even damage the engine.

- **Park it fast** – Park in the first available parking space you come to. Don't waste a lot of gas – and time – circling the lot looking for a better (and perhaps nonexistent) place to park.

- **Use shades** – Reflecting sun shades made especially for cars, vans, and trucks can be spread inside the front windshield and above the dashboard to reflect heat away from the car and help keep it cool when it's parked, especially if you haven't been able to park in the shade.

- **Cool off before you turn the AC on** – You'll use less gas if you roll the windows down and let a hot car cool off a little before cranking up the air conditioning.

- **Use your smaller car most often** – If you happen to have two cars, drive the smaller, more fuel-efficient car most frequently, especially on shorter trips and for doing errands.

> FACE THE FACTS: More than half the oil used to produce the gasoline you put in your tank is imported. The U.S. consumes about 20 million barrels of oil each day, two-thirds of which is used for transportation. Petroleum imports cost us about $2 billion a week. That's money that could otherwise fuel our own economy. *Source: U.S. Department of Energy*

❖2❖
DRIVE CHEAP

No one ever won a prize for spending the most money possible on gas.

On the contrary, your goal should be to spend the least amount of money – and use the least amount of gas – you can to cover the distance you want to go.

Here's how:

- **Buy the cheapest gas you can find** – People who pride themselves on being "bargain hunters" at the shopping mall can go bargain hunting for gasoline just as easily. Is the "hunt" worth the payoff? In my Silver Spring, Maryland, neighborhood, regular unleaded gas that was selling for $3.15 a gallon was costing drivers $3.35 a gallon just a few miles away . . . on the same day! To find the station in your area with the lowest prices, go to www.gasbuddy.com or www.gaspricewatch.com. Since today's gasolines are all very similar in content, you can shop for price without worrying you'll impact your vehicle if you change from brand to brand.

- **Use the right octane gas** – High-octane fuel sells for as much as 20 cents a gallon more than regular unleaded. But unless you're driving a Maserati, you probably don't need to pay the premium. Less than 10 percent of all cars sport the high-compression engines that require the more expensive fuel. Even so, as many as 30 percent of Americans fill up with high octane. Don't waste your money. Check your owner's manual to find out what octane gas your vehicle's engine needs. Skip premium gas if you don't need it, and pocket the savings.

- **Don't top off your tank** – Stop filling up your gas tank as soon as you hear the nozzle on the gas pump click. Otherwise, you risk overflowing your tank – wasting a lot of money on fuel that waters the pavement rather than helps move your vehicle on down the road.

- **Use gas discounters** – Many retailers not traditionally thought of as gas stations have installed pumps to offer their customers gasoline savings that range between 3 and 7 cents a gallon. Even Kroger offers gas at 536 food stores. Regional discount gas station chains like QuikTrip and Love's are also popping up around the country. The gasoline at Costco is the same fuel that name brand-stations like Shell, Chevron, and BP typically sell. Often it comes from the same refineries. The branded fuel contains a small amount of detergent that can help clean important working parts of an engine, such as the valves and fuel injectors. But even the lowest-price gasoline meets detergent standards set by the Environmental Protection

Agency in 1995. Thrifty consumers can buy extra detergent at auto-parts stores for as little as 99 cents and simply pour it into the tank every 3,000 miles or so for a considerable savings over the cost of branded midgrade or premium gasoline. Alternatively, fill up occasionally on a name-brand gasoline that contains extra cleansing compounds and buy discount gas the rest of the time to keep valves and fuel injectors from clogging, says Dennis DeCota, executive director of the California Service Station and Automotive Repair Association.

- **Get the rebate** – Various gas rebate cards are available to help you receive anywhere from a 2-10 percent credit on your bill when you purchase gas. Several of the cards require you to buy gas from a specific gasoline vendor, like Shell. Others can be used at any station. Perhaps the most desirable cards offer cash back on all gas purchases, regardless of the station or brand you use. While most gas rebate cards do not require annual fees, they may impose high interest rates. Many consumers use them only to buy gas and pay off the balance promptly to avoid any extra charges. Search for more information about gas rebate cards at www.google.com or www.yahoo.com.

- **Use fuel discount offers** – Some grocery stores offer coupons and special promotions that discount gas when you purchase groceries and other items. Stores may discount gas if customers pay with the store's own discount card. Kroger, for example, was giving customers a 10-cents-a-gallon discount for every $100 in groceries they buy each month.

- **Pay cash** – Occasionally, gas will be cheaper if you pay for it with cash. The key is to get to know the gas vendors in your neighborhood and the kind of deals they offer to their customers. Find the best bargain you can so you won't spend unneeded dollars on fuel.

When It Comes to Engine Oil . . .

- **Use "Energy Conserving" oil** – You can improve your gas mileage by 1-2 percent by using the manufacturer's recommended grade of motor oil in your engine. Look for motor oil that says "Energy Conserving" on the API performance symbol to be sure it contains friction-reducing additives.

- **Change oil regularly** – Follow the oil change recommendations of your vehicle manual to keep your engine running as efficiently as possible. Regular oil and filter changes should be considered every 3,000 miles or every 3-6 months.

> FACE THE FACTS: It took more than 200 million years to form all of the oil beneath the surface of the earth. It has taken 200 years to consume half that amount. If current rates of consumption continue, the world's remaining resources of conventional oil could be consumed in the foreseeable future.

✧3✧
DRIVE LESS

Millions of Americans have already figured out how to cut back on the amount of gas they use simply by cutting back on how much they themselves drive. Follow their lead:

- **Walk** – If you're traveling distances of less than a mile and don't have to carry heavy loads, try walking instead of driving. An added benefit? Hoofing it will help keep you in good physical shape. HINT: carry a small backpack or shoulder bag for items you need to tote along.

- **Bicycle** –You can travel hundreds of miles by bicycle on no more than a thimbleful of oil. More than half of all commuting trips are 5 miles or less in length, a distance that could easily be covered by bicycle. So park your four-wheeler and pull out the two-wheeler. If you need to, get a basket for the front and a rack for the back to make carrying light loads easier.

- **Scoot** – That's right, scoot! Electric scooters (not gas-burning ones) use rechargeable batteries that can plug right into an electrical socket.

They're great for short commutes, and you won't have a parking hassle when you arrive at your destination; you can just take your scooter into the office with you or lock it to a parking meter the way you would a bicycle. Some scooters come with seats and grocery baskets to make shopping and doing errands easier. For more information, check out www.electric-scooters.com.

- **Moped** – Electric mopeds feature battery packs that cost only 15 cents to recharge and will let you travel up to 25 mph, covering a distance of 20 miles before they need to be recharged. For more information, Google "moped" on the Internet and search for the vehicle with the lowest emissions ratings.

- **Use Flex Cars and Zip Cars** – Flex Cars and Zip Cars make it easy for drivers to borrow a vehicle when they need one in lieu of owning their own. For the price of a membership, drivers get access to vehicles that are located near subway stations or bus stops; insurance; gas; and often guaranteed parking places. For more information, see www.zipcar.com or www.flexcar.com.

- **Combine trips** – Plan ahead, so that you combine several errands into one trip. You'll save time as well as money. Several short trips taken from a cold-starting engine can use twice as much fuel as a longer multipurpose trip covering the same distance when the engine is warm.

- **Avoid the rush** – Sitting in a traffic jam is one of the worst ways to waste gas. Whenever possible, stagger your work, travel, or errand schedule to avoid peak rush hours. (If you are stuck in traffic, turn the car off or at least shift into neutral; both measures will save gas.)

- **Find another way** – Choose alternatives to the most densely traveled thoroughfares. It might be faster, safer, and more gas-efficient to get off the highway and take side roads to your destination.

- **Pay your bills online** – Online banking was originally designed to save people time and help them avoid late fees; it turns out that electronic bill-paying is a great way to save gas, as well. Several software programs are available to help you set up an online check-writing process. If you haven't done so already, arrange to have your paycheck direct-deposited into your bank account by your employer so you don't have to drive to a bank or drive-through teller. Find an ATM machine within walking distance to get cash "gasoline free" when you need it.

- **Change addresses** – Some people actually choose the location of their residence based on the availability of mass transit options. They may not have a car or may want to sell it if they do; or they may simply want to minimize the amount they drive. Urban areas usually have better mass transit systems than suburban and rural areas, though newer communities are doing a better job of integrating energy-saving transportation options into their plans.

Use the Internet

If you're not using the Internet to keep your car parked, you're doing more to put money in the pockets of the oil companies than in your own.

Here's how your computer can become your own personal oil gusher:

- **Shop online** – You can buy almost any product via the World Wide Web, from clothing and sporting equipment to books or the latest DVD. Take advantage of free home delivery services to shop from the fuel-efficient comfort provided by your computer. After all, why drive over to the local movie rental outlet when you can order DVDs online more cheaply and they get delivered to your house by mail for free? (NOTE: To maximize fuel-efficiency, always choose the "standard" delivery option, rather than express or expedited.)

- **That includes food** – Many grocery stores now deliver online orders. The service saves customers not just the hassle of driving to the store but the time and gas involved in circling the parking lot looking for a place to park, a benefit that seems well worth the modest delivery fee the stores charge. Customers who use these home deliveries to stock up on cheap bulk items like canned goods and paper products find that a monthly online delivery service saves them several small gas-guzzling trips to the store. Meanwhile, the stores design efficient delivery routes to maximize energy savings on their end.

- **Shop at night** – If you have to go shopping or to the mall, try going at night or during off hours, when traffic will be minimal and you'll be more likely to find a parking spot quickly when you arrive.

- **Walk to most shops** – Try to shop at locations that allow you to find most of your purchases in one place. It is more fuel-efficient to park and shop for several hours in one location rather than to drive to several locations.

- **Use the Internet or the telephone to research before you buy** – Even if you need to go to a store, research your options before you get in the car. Consumers waste a tremendous amount of gas driving to places that don't have what they need simply because they didn't check out the place of business first.

FACE THE FACTS: In the past, dependence on oil has shaken our economy severely. Oil price shocks and price manipulation by the OPEC cartel from 1979 to 2000 cost the U.S. economy about $7 trillion, almost as much as we spent on national defense over the same time period and more than the interest payments on the national debt. Each major price shock of the past three decades was followed by an economic recession in the United States. With growing U.S. imports and increasing world dependence on OPEC oil, future price shocks are possible and would undermine the U.S. economy.

4
DRIVE IN TUNE

Keep your car "singing" to get the most miles out of every gallon of gas.

- **Get a tune-up** – Improve your gas mileage by an average of 4.1 percent, depending on the tune-up and how seriously your vehicle should be serviced. Check your owner's manual for recommendations on how frequently your vehicle should get tuned up. New cars may need a tune-up after the first 30,000 or 50,000 miles they're driven. Every car should be tuned up every year.

- **Look at the oxygen sensor** – Ask your mechanic to take a special look at your engine's oxygen sensor, the master switch in the engine's fuel control feedback loop. Repairing a faulty oxygen sensor could benefit your gas mileage as much as 40 percent.

- **Check and replace air filters regularly** – Replacing a clogged air filter can improve your car's gas mileage by 10 percent. A clean air filter reduces damage to the inside of your engine, too.

- **Change oil regularly** – Follow the oil change recommendations of your vehicle manual to keep your engine running as efficiently as possible. You can improve your gas mileage by 1-2 percent by using the manufacturer's recommended grade

of motor oil. As noted before, look for motor oil that says "Energy Conserving" on the API performance symbol to be sure it contains friction-reducing additives.

Pump'em Up

- **Keep tires properly inflated** – Buy a tire pressure gauge and use it often to check tire pressure. Gas mileage will improve by around 3.3 percent if you keep your tires properly inflated. Under-inflated tires can lower gas mileage by 0.4 percent for every 1psi drop in pressure of all four tires. Check  your owner's manual for appropriate inflation levels. This information is also usually available on the jamb of the driver's-side door.

- **Replace old tires with the same make and model as the original ones** – Replacement tires should be the same make and model as the tires that were on your vehicle when it was new in order to maintain maximum fuel efficiency.

Pay attention to suspension – Vehicle suspension misalignment can lead to poor fuel economy. All four tires should be checked for abnormal or premature tire wear. Replacing the tires, rotating tires, or doing a front-end alignment could all improve fuel economy.

Tips for the Two-Car Family

If your family owns two cars, chances are one of them is a minivan, SUV, or wagon. Hopefully, the other is a compact, mid-size sedan, or maybe even a hybrid. In all likelihood, one vehicle gets better gas mileage than the other, yet you need both to meet your family's needs. Consider these suggestions to maximize gas savings:

- **Drive the most fuel-efficient vehicle whenever possible** – Don't choose the gas-wasting SUV if you're simply running up to the post office; on the other hand, it'll be tough to cram half the soccer team into the small but energy-efficient compact. Choose the right vehicle to meet your driving needs, always trying to drive your fuel-efficient vehicle most of the time.

- **Maintain all vehicles to maximize fuel efficiency** – All vehicles need to be given equal treatment when it comes to engine maintenance, tire inflation, and overall upkeep.

- **Carpool with other families** – If you have a larger vehicle, take advantage of its passenger capacity. Carpool with other parents so that you minimize trips to sporting and school events.

- **Resist the urge to use more than one vehicle to go to the same event** – Drive the more efficient vehicle as the primary mode of transportation for the entire family.

- **Walk, bicycle, and use mass transit** – No matter how many vehicles you own, you should still consider walking if the distance is manageable and you don't need to carry a heavy load. Kids and adults alike will benefit from the exercise they'll get from walking or bicycling. And wherever mass transit is available, use it. You'll save money on gas, parking, and vehicle wear and tear.

- **Buy the most fuel-efficient vehicle to meet your family's needs** – When you're in the market for a new vehicle, put fuel efficiency right up there with cup holders as a desirable option. You can compare energy-efficiency ratings for vehicles in all sizes at www.fueleconomy.gov. See the fuel cost calculator on page 47 to help you anticipate potential fuel costs given today's prices at the pump.

HOW MUCH WILL YOU SAVE?

Let's assume you:

- Drive 15,000 miles per year
- Your car gets 21 miles per gallon
- Gas costs $3.00/gallon
- You spend $2,142/year on gasoline

If you . . .
- Drive sensibly at highway speeds, you can increase fuel efficiency 33 percent and save $532/year ($44.33/month).

- Observe the speed limit, you can increase fuel efficiency 23 percent and save $401/year ($33.40/month).

- Change your vehicle's air filters, you can increase fuel efficiency 10 percent and save $195/year ($16.25/month).

- Tune up your vehicle, you can increase fuel efficiency 4.1 percent and save $85/year ($7.04/month).

- Replace a faulty oxygen sensor, you can increase fuel efficiency 40 percent and save $612/year ($51.03/month).

- Keep your vehicle's tires properly inflated, you can increase fuel efficiency 3 percent and save $62.55/year ($5.21/month).

✶5✶
DRIVE A GAS STRETCHER, NOT A GAS GUZZLER

Given these days of high gasoline prices, uncertain oil supplies, and the global warming associated with burning fossil fuels, it makes sense to stretch every gallon of gasoline just as far as it will go. You won't be able to do that in a Hummer H2, which only gets around 10-13 miles per gallon (mpg). Even SUVs, light trucks, and minivans only average 21 mpg, while family cars range around 27.5 mpg. We can – and must – do better. Many economists, oil industry analysts, and environmentalists agree that fuel economy standards need to be increased to over 40 mpg by 2015, and 55 mpg by 2025.

We can achieve that goal by passing new Corporate Average Fuel Economy (CAFE) standards. The Congressionally mandated CAFE standards initially improved new car and truck fuel economy by 70 percent between 1975 and 1988. CAFE saved American consumers at least $92 billion, reduced gasoline use by 60 billion gallons, and kept 720 million tons of global warming pollutants out of the atmosphere.

But the standards haven't been updated since 1985. Increasing fuel efficiency to 40 mpg in the next decade could cut annual greenhouse gas emissions by 106 million tons, reduce our dependence on oil from terror-prone and environmentally sensitive regions, and help consumers realize annual gasoline savings of as much as $50 billion.

What can you do?

- **Drive your most fuel-efficient vehicle** – If you have two vehicles, as often as possible drive the one that stretches the most miles out of every gallon of gasoline.

- **Support increasing CAFE standards** – Encourage local and national officials to raise fuel efficiency standards for cars, minivans, light trucks, and Hummers. According to the American Council for an Energy Efficient Economy, raising CAFE standards by 5 percent annually until 2012 and by 3 percent per year thereafter could save 1.5 million barrels of oil per day (MBD) by 2010, 4.7 MBD by 2020, and 67 billion barrels of oil MBD over the next 40 years. This is 10-20 times greater than the potential oil supply from the Arctic National Wildlife Refuge.

- **Remember: A gas-stretching vehicle will get you to the same location as a gas guzzler, but at far less economic and environmental cost** – Of the vehicles you own, if you don't know which one gets better mileage, compare actual gas consumption and miles traveled in each vehicle over the course of a month. Compare highway vs. city driving; most vehicles achieve higher mileage rates under highway conditions.

- **Make your new car a gas stretcher** – When you're in the market for a new set of wheels, buy the most fuel-efficient model you can afford. You'll find additional new car buying tips in this book, in the chapter titled "What to Consider If You're Buying a New Vehicle."

FACE THE FACTS: U.S. consumers currently send over $200,000 every minute overseas to buy oil.
Source: Union of Concerned Scientists

❖6❖
DON'T DRIVE ALONE

The benefits of sharing a ride extend beyond the gas gauge and your bank account. Commuting with others does wonders to limit traffic congestion, improve air quality and benefit human health. Here are three popular options.

- **Carpool** – You can cut your weekly fuel costs in half and save wear on your car if you take turns driving with other commuters. You might also get to your destination faster if you're driving with others. Many urban areas allow vehicles with multiple passengers to use special High Occupancy Vehicle (HOV) lanes to bypass clogged, gas-burning freeway arteries.

- **Share the ride** – If you can't find anyone to carpool with, check your local yellow pages or the Internet for information about "Ride Share" programs that will hook you up with other commuters who have approximately the same destination you do and who may be happy to share the ride – and the costs of gasoline. Visit the website www.AlterNetRides.com to find a wide variety of ride sharing opportunities.

- **Get on the bus** – Buses, subways, and other public transit systems offer another smart alternative to driving your own car. The American Public Transit Transportation Association provides information about public transportation options available in your state. For more information, check out www.apta.com/links/state_local.

WHEN YOU DO GAS UP...

- **Be an early bird or a night owl** – Buy gasoline during the coolest time of the day, either in early morning or late evening. Because gasoline is denser at cooler temperatures, you'll get more for your money.

- **Don't top off your tank** – Stop filling up your gas tank as soon as you hear the nozzle click. Overfilling the tank can lead to overflow and air pollution problems as well as waste your money.

Non-Gasoline Fuel Options

Gasoline derived from petroleum is not the only energy source available to fuel our cars, trucks, and other vehicles. Some alternative fuels, like ethanol, have been around for decades. Others, like fuel made from vegetable oil, are just beginning to find their place in vehicle engines.

- **Explore alternatives to gasoline** – An increasing number of alternative fuel vehicles (AFVs) are available from both domestic and foreign manufacturers. AFVs run on natural gas, propane, electricity, ethanol or E85, a mixture of 85 percent ethanol and 15 percent gasoline.

- **Consider all the options** – There are three basic types of AFVs: flex-fuel, dual-fuel, and dedicated vehicles. A flex-fuel vehicle has one tank and can accept any mixture of gasoline and either methanol or ethanol. It's estimated that 1 to 2 million flex-fuel vehicles are already on American roads. Dual-fuel vehicles have two tanks: one for gasoline, and one for either natural gas or propane. The vehicles can switch between the two fuels. Dedicated vehicles are designed to be fueled only by an alternative fuel. For more information on all of these options, surf over to www.eere.energy.gov/cleancities/vbg/consumers/choices.shtml.

- **Put ATVs on the list** – ATVs – alternative technology vehicles – include hybrid electric vehicles (see the hybrid section below), neighborhood electric vehicles, and even electric bicycles. You can learn more about them at www.eere.energy.gov/consumerinfo/energysmart_vehicles.html.

- **Look into biodiesel** – Some cars are being retrofitted to use biodiesel, which combines diesel fuel and vegetable oil – or even uses vegetable oil alone. Check out German-made engine conversion kits at www.elsbett.com if you're interested in using vegetable oil instead of gasoline to drive to work or the mall.

- **Support hydrogen-powered fuel cells** – Fuel cells run on the energy generated when hydrogen and oxygen are mixed. Vehicles powered by fuel cells will use one-third the energy of today's cars – and none of it from oil! For now, the cost of developing fuel cells and vehicles is extremely high. More financial incentives are needed to ramp up research and development to get production to 100,000 vehicles by 2010 and even 2.5 million by 2020. Let auto manufacturers, elected officials, and federal agencies know you support accelerating the rate at which hydrogen-powered fuel cell vehicles reach the road. Start by contacting your member of Congress at www.house.gov.

FYI:

Ethanol – Ethanol is the most widely used alternative transportation fuel. It is an alcohol typically made from corn or corn by-products, using a process similar to brewing beer. Vehicles that run on ethanol generate less pollution than traditional vehicles. Some flexible-fuel vehicles can use as much as 85 percent ethanol, really stretching out gasoline supplies.

Methanol – Another alcohol-based fuel, methanol is usually produced from natural gas. It can also be produced from plant or animal waste.

Biodiesel – Biodiesel can be made from several types of oils, such as vegetable oils and animal fats. Each year about 30 million gallons of biodiesel are produced in the U.S. from recycled cooking oils and soybean oil. Biodiesel is typically used as a blend, though increasingly, consumers are converting their own car engines so that they can use biodiesel almost exclusively. They're getting the oil from Chinese restaurants and fast-food establishments that cook a lot of French fries!

SAVING GAS ON THE JOB

You probably spend a substantial portion of your gasoline budget just commuting back and forth to work. Here are a few ideas to save gas and money and still get the job done.

- **Don't go to work** – Well, I don't mean you should quit your job. Just look for ways to work from home as much as possible to minimize the expense of driving to work. If you have a computer, phone, and fax machine, can you telecommute one or two days a week? Meet with your supervisor or personnel manager to determine what company policies will help you cut back the amount of driving you do for your job.

- **Go to work at a different time** – If you need to keep a five-day-a-week schedule, can you alter your hours to avoid driving during rush hour?

Stop-and-go traffic burns up gasoline almost as fast as lighting a match to it. Check on the possibility of staggering your work hours to avoid the biggest traffic jams.

- **Carpool** – Yes, carpooling has been around for a long time. That's because it still makes sense to share not just the expense of driving but the hassle. Plus, in most cities, carpools get access to faster, "high occupancy vehicle" (HOV) lanes that single drivers don't. You can contact your city's metropolitan transit authority to locate carpools in your neighborhood. You can also use community e-mail listservs and bulletin boards to network with other riders with whom you share a common destination.

- **Bike to work** – Many offices now provide lockers and showers for employees who bicycle, walk, or jog to work.

- **Use mass transit** – Even when they have to buy a bus or subway ticket, many commuters find that using mass transit is cheaper than driving to work and paying for parking. Some employers subsidize their employees' use of mass transit with programs like Metrochek. Metrochek is a farecard voucher program provided as an employee benefit by more than 2,500 public and private employers in the Washington, D.C., metropolitan area, including the federal government. The Metrochek farecards are accepted by more than 100 bus, rail, and vanpool commuter services in the region, and can be used as vouchers when purchasing fares for other transit services. The benefit works just like many other commonly available fringe benefits such as employer-provided health insurance. Employees are not taxed for the value of the Metrocheks they receive, while employers can deduct the cost of providing the program as a business expense. The monthly Metrochek benefit can be any amount the employer chooses to provide, although a maximum of $105 per month is allowable tax-free or pre-tax to employees.

FACE THE FACTS: If you carpool or take mass transit to work, you'll not only hear more money jangling in your pocket. You'll be making a contribution to a healthier planet. Leaving your car at home just two days a week will reduce the carbon dioxide emissions that cause global warming by 1,590 pounds per year.

Vacation and Traveling Tips

You won't enjoy your vacation nearly as much if you have to worry about how much the gas to get you there is going to cost. These ideas will help you save several dollars every time you gas up.

- **Get a quick tune-up before you leave** – Have your car's oil, transmission, and spark plugs checked to improve gas mileage throughout your road trip.

- **Don't start the engine until you're ready to go** – You can waste a lot of gas idling while everyone else is still packing up last-minute items for the trip. Wait until all your passengers are in your vehicle and then turn on the ignition.

- **Be trip smart and traffic savvy** – Plan your route to bypass as much traffic congestion as possible. Avoid rush hour in the cities you're traveling through. Use www.mapquest.com to seek out the shortest distance possible to your destination.

- **Walk when you get there** – Even if you're driving to your destination, try to leave your car in "park" once you arrive. Choose vacation destinations that have plenty of options for pedestrians or benefit from mass transit so you don't need to drive everywhere to enjoy yourself.

- **Put away the roof rack** – A roof rack or carrier provides additional cargo space and may allow you to meet your traveling needs with a smaller, gas-saving car. However, a loaded roof rack can decrease your fuel economy by 5 percent. Reduce aerodynamic drag and improve fuel efficiency by placing items inside the trunk whenever possible. Don't leave the roof carrier on permanently; as soon as your trip is over, remove it.

- **Lighten your load** – An extra 100 pounds in the trunk reduces a typical car's fuel efficiency by 1-2 percent. So lighten your load, and avoid carrying unneeded items, especially heavy ones.

- **Rent a green car** – If you have to rent a car when you reach your destination, opt for the most fuel-efficient model available. Many auto rental companies now offer their customers Toyota Prius and Honda Civic hybrids. Companies in Hawaii make available a "Bio-Beetle" Volkswagen that runs on vegetable oil. Companies like EV Rental Cars offer electric vehicles for rent.

- **Enjoy greener service** – If you live in the Boston, Massachusetts, area, you can contact PlanetTran (www.planettran.com) for airport transportation via a hybrid vehicle. The Better World Club (www.triplee.com) provides 24-hour nationwide roadside assistance for bicyclists as well as motorists. It also offers its members 10 percent discounts on hybrid and electric car rentals.

WHAT TO CONSIDER IF YOU'RE BUYING A NEW VEHICLE

First, ask yourself a series of questions. Do you need this vehicle to move one or two people, or a large group? Are you going to use it primarily for commuting to work, for doing errands around town, or for covering long distances? What is the most fuel-efficient vehicle you can consider, given your driving needs and budget?

Many vehicle models come in a range of engine sizes and trim lines, resulting in different fuel economy values. When purchasing a vehicle, check with the manufacturer on projected fuel economy for that specific model. Similar ratings are available for used cars.

Remember that options like four-wheel drive and third-row seats increase the weight your vehicle carries, adding to its fuel consumption. If you don't need the extras, skip them. You'll be happy you did when you go to fill up your tank.

Listed below are the U.S. Environmental Protection Agency's Model Year 2005 Fuel Economy Leaders, the vehicles with the highest fuel economy in the most popular classes. The list includes vehicles with both automatic and manual transmissions; note that manual transmissions are always more fuel-efficient.

EPA's Model Year 2005 Fuel Economy Leaders (mpg city/hwy)

Two-Seater
- Honda Insight manual 61/66; automatic 57/56

Minicompact
- Mini Cooper manual 28/36; automatic 26/34

Subcompact
- Volkswagen New Beetle (diesel) manual 38/46
- Volkswagen New Beetle (diesel) automatic 36/42

Compact
- Honda Civic Hybrid manual 46/51; automatic 48/47

Midsize
- Toyota Prius Hybrid automatic 60/51
- Hyundai Elantra manual 27/34

Large Cars
- Toyota Avalon automatic 22/31

Small Station Wagon
- Volkswagen Jetta (diesel) manual 36/43

- Volkswagen Jetta (diesel) automatic 32/43

Midsize Station Wagon
- Volkswagen Passat Wagon (diesel) automatic 27/38
- Ford Focus manual 26/35; automatic 26/32

Cargo Vans
- Chevrolet Astro 2WD automatic 16/22
- GMC Safari 2WD automatic 16/22

Minivans
- Honda Odyssey 2WD automatic 20/28

Passenger Vans
- Chevrolet Astro 2WD automatic 16/21
- GMC Safari 2WD automatic 16/21

SUV
- Ford Escape Hybrid HEV 2WD automatic 36/31
- Toyota RAV4 2WD manual 24/30

Standard Pickup Trucks
- Ford Ranger Pickup 2WD manual 24/29; automatic 22/26
- Mazda B2300 2WD manual 24/29; automatic 22/26

NOTE: The Ford Escape Hybrid SUV gets substantially better mileage than the Honda Odyssey minivan.

Calculate Your Fuel Costs:

The federal government has developed the Fuel Cost Calculator to help you anticipate annual fuel costs, as well as compare gas and money savings between vehicles. It makes clear how much gas and money you can save simply by driving a more fuel-efficient vehicle. You can visit www.fueleconomy.gov/feg/savemoney.shtml to download your own fuel cost calculator as you look for fuel and cost economy.

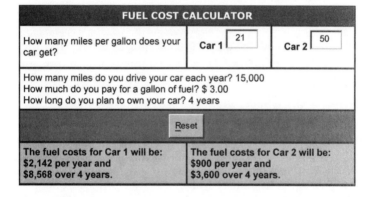

TRY A HYBRID

In case you haven't noticed, the hybrids, like the Toyota Prius and the Honda Insight, really stand out for their fuel efficiency. They're sporty and fun to drive, too. In fact, the 2004 Toyota Prius was *Motor Trend* magazine's Car of the Year, beating out dozens of other conventionally fueled vehicles for this coveted award due to its "performance, engineering, and overall significance." In bestowing its award, the editors of *Motor Trend* described the Prius as a "capable, comfortable, fun-to-drive car that just happened to get spectacular fuel economy." Not bad for a car that achieves 60 mpg!

Don't let the hybrid technology scare you off. Hybrids simply combine a smaller gasoline engine with a battery-powered electric motor. In doing so, they double the mileage of conventional cars. Hybrids burn little fuel as they slow down, and can come to a complete stop when waiting in traffic. You recharge their batteries every time you hit the brakes. You don't have to plug them into an electrical socket; they use the same gasoline other cars do, though less of it.

Tax Incentives

Another advantage of buying a hybrid: You can garner a $2,000 federal "clean fuel" tax deduction if you make the purchase by the close of 2005. Even if you purchased the vehicle before 2005, you can claim the deduction by filing an amended tax return for the tax year in which the vehicle was purchased. For your vehicle to qualify, you must purchase the vehicle new and for your own use, not for resale. You must drive it mostly in the United States. In addition, the vehicle must meet all federal and state emissions regulations.

Eligible vehicles include:

Ford Escape Hybrid – Model Year 2005
Honda Accord Hybrid – Model Year 2005
Honda Civic Hybrid – Model Year 2003-2005
Honda Insight – Model Year 2000-2005
Lexus RX 400h – Model Year 2006
Toyota Highlander – Model Year 2006
Toyota Prius – Model Year 2001-2005

More information about federal tax deductions can be found at www.fueleconomy.gov/feg/tax_hybrid.shtml.

NOTE: Several states also offer tax deductions when you purchase a hybrid. Check state tax laws when you buy your hybrid to enjoy the full tax benefit of this investment.

For more information about the Toyota Prius, visit www.toyota.com.

For more information about the Honda Insight, EPA's top rated hybrid, visit www.hondacars.com.

If you need an SUV, take a look at the new Ford Escape hybrid. www.fordvehicles.com.

More Car-Buying Tips

These additional car-buying tips will make filling up at the pump less painful:

- **Consider diesel** – Diesel engines typically travel about 30 percent farther on a gallon of gas than a similar-sized gasoline engine. Consequently, they generate less carbon dioxide, which has been linked to global warming. Unfortunately, they do produce more of the particulates that cause smog. Starting in 2006, however, the federal government is requiring diesel fuel to reduce its sulfur content, which will make it 90 percent cleaner to burn than it is today. Meanwhile, Volkswagen and DaimlerChrysler are selling diesel cars in the U.S. that are recommended by the EPA for their low gas mileage.

- **Look under the hood** – Engines featuring 4-valve cylinders and variable valve timing are the most efficient. You can find them in most cars and smaller SUVs, but not most medium and large pickup trucks and larger SUVs.

- **Seek out something sleek** – Aerodynamically designed vehicles reduce drag, saving fuel. And cars and trucks made of aluminum and high-strength steel can reduce vehicle weight while maintaining size and strength as well as improving safety.

- **Ask for automatic shut-off** – New advances in automotive technology are leading to engines that can shut off at stoplights, then restart instantly as needed. They'll also shut down extra cylinders when the car is cruising. You'll already find "auto-stop" on the lot; cylinder deactivation (or "displacement on demand") will soon be available from General Motors.

- **Vary your transmission** – Enhanced transmissions offer another option for improving gas mileage. Five- and six-speed transmissions provide better fuel economy, while continuously variable transmissions do away with gears altogether. Some model options? The Ford Explorer uses a five-speed automatic transmission; the Honda Civic HX, Audi, and GM Saturn Vue SUV offer continuously variable transmissions. Alternatively, specify overdrive transmission. When you use overdrive gearing, your car's engine speed goes down, saving gas and reducing engine wear.

For More Information

www.fueleconomy.gov
This website lists gas mileage tips and a fuel economy guide offering gas mileage estimates and more information for 1985-2005 model year cars. You can compare cars to find the most fuel-efficient vehicle that meets your needs.

www.eere.energy.gov/cleancities/vbg/consumers/choices.shtml
This website provides a Vehicle Buyer's Guide for consumers in the market for a clean and efficient vehicle. It describes several models of alternative fuel vehicles and alternative technology vehicles available from both domestic and foreign manufacturers. The site is operated by the Clean Cities program of the U.S. Department of Energy's Energy Efficiency and Renewable Energy Office.

www.aceee.org/transportation/topics.htm
The American Council for an Energy Efficient Economy is a nonprofit organization dedicated to advancing energy efficiency as a means of promoting both economic prosperity and environmental protection. These web pages offer a comprehensive look at various transportation alternatives.

www.rmi.org
The Rocky Mountain Institute is a nonprofit research center that does some of the most creative thinking

in the country on how to solve national and global energy problems. Their website explains practical and economical approaches to solving the current oil crisis.

www.AlterNetRides.com
This website helps people share rides from one destination to another across the United States.

www.gaspricewatch.com and www.gasbuddy.com
Both websites will help consumers locate the cheapest gas in their driving range.

www.aaa.com
The American Automobile Association offers fuel conservation tips and regular updates on current regional and national gas prices.

www.flexcar.com and www.zipcar.com
Both websites provide information for consumers who are interested in sharing a car with other drivers rather than owning their own vehicle outright.

In Canada:

http://oee.nrcan.gc.ca/transportation/personal-vehicles-initiative.cfm
The Personal Vehicles Initiative provides Canadian motorists with helpful tips on buying, driving, and maintaining their vehicles to reduce fuel consumption and greenhouse gas emissions that contribute to climate change. This website offers an

alternative fuels vehicle guide, tips to reduce idling, vehicle comparison charts, and gas consumption calculators.

The Top Ten Ways to Beat the High Price of Gas

1. **Drive smart** – Avoid quick starts and stops, use cruise control on the highway, and don't idle.

2. **Drive the speed limit** – Remember – every 5 mph you drive above 60 mph is like paying an additional $0.10 per gallon for gas.

3. **Drive less** – Walk, bicycle, use a scooter or moped, combine trips, and telecommute.

4. **Drive a more fuel-efficient car** – Consider one of the new hybrids; at the very least, choose from among the EPA's "Fuel Economy Leaders" in the class vehicle you're considering.

5. **Keep your engine tuned up** – Improve gas mileage by an average of 4.1 percent by maintaining your vehicle in top condition.

6. **Carpool** – According to the Natural Resources Defense Council, 32 million gallons of gasoline would be saved each day if every car carried just one more passenger on its daily commute.

7. **Use mass transit and "Ride Share" programs** – Why pay for gasoline at all?

8. **Keep tires properly inflated** – Improve gas mileage by around 3.3 percent by keeping your

tires inflated to the proper pressure. Replace worn tires with the same make and model as the originals.

9. **Buy the cheapest gas you can find** – Buy gas in the morning, from wholesale shopper's clubs, and using gasoline rebate cards. Track neighborhood prices on the Internet.

10. **Support higher CAFE standards and the development of alternative fuels** – Ultimately, our best hope for beating the gas crisis is to increase fuel efficiency while we transition to renewable and non-petroleum based fuels. Endorse efforts to boost average fuel efficiency to at least 40 mpg. Support programs that promote research and development of alternatives to transportation systems based on oil.

About the Author

Diane MacEachern is a best-selling author, successful businesswoman, and environmental entrepreneur. She is also the founder and CEO of The World Women Want, an environmental clearinghouse that helps women, their children, and their families live healthier, safer, and more rewarding lives (www.theworldwomenwant.com).

Her first book, *Save Our Planet: 750 Everyday Ways You Can Help Clean Up the Earth*, was featured on national television programs ranging from CNN Headline News to *Live with Regis and Kathie Lee* to *The ABC Network Television Earth Day Special*. Ms. MacEachern's weekly newspaper column, *Tips for Planet Earth,* was nationally syndicated by the Washington Post Writers Group. She has also written for *Good Housekeeping*, *Family Circle*, *Self,* the *Christian Science Monitor*, *Ladies Home Journal*, and the *Baltimore Sun.*

Working with the U.S. Environmental Protection Agency, Ms. MacEachern helped educate the public about the causes of global warming and climate change. She has also helped non-profit organizations advance new, efficient sources of energy. She is a frequent speaker at conferences, workshops, and seminars.

Ms. MacEachern received a master's degree from the School of Natural Resources and Environment at the University of Michigan. She lives with her family in the energy-efficient home they built in Maryland and drives a gas-saving hybrid vehicle.